Naked Mole-Rats

by Gail Jarrow and Paul Sherman

A Carolrhoda Nature Watch Book

Carolrhoda Books, Inc./Minneapolis

For Robert
    —G.J.

For Cindy, Philip, Laura, and Peter
Sherman
                    —P.S.

CHILDRENS ROOM

Text copyright © 1996 by Gail Jarrow and Paul Sherman
Illustrations pp. 6, 9, 23 by Laura Westlund copyright ©
1996 by Carolrhoda Books, Inc. All photographs courtesy
of Raymond A. Mendez except: pp. 5, 14, 27, 28–29, 33,
35, 39, 40–41, 42, Jennifer U. M. Jarvis; p. 48 (left),
Robert Jarrow.

This book is available in two editions:
Library binding by Carolrhoda Books, Inc.
Soft cover by First Avenue Editions
c/o The Lerner Group
241 First Avenue North, Minneapolis, MN 55401

LIBRARY OF CONGRESS CATALOGING-IN-PUBLICATION DATA

Jarrow, Gail.
    Naked mole-rats / by Gail Jarrow and Paul Sherman.
      p.   cm.
    "A Carolrhoda nature watch book."
    Includes index.
    Summary: Describes this strange-looking rodent which
lives a social life in a system of tunnels under the
brick-hard soil of East Africa.
    ISBN 0-87614-995-6 (lib. bdg.)
    ISBN 1-57505-028-5 (pbk.)
    1. Naked mole rat—Juvenile literature. [1. Naked mole
rat. 2. Rodents.] I. Sherman Paul W., 1949–
II. Title.
QL737.R628J365   1996
599.32'34—dc20   95–44979

Manufactured in the United States of America
1 2 3 4 5 6 – JR – 01 00 99 98 97 96

CONTENTS

THE UNDERGROUND MYSTERY

A mysterious creature lives under the brick-hard soil of East Africa. It is a strange-looking beast. You might even call it ugly. It has a skinny tail and wrinkly skin, huge front teeth and tiny eyes.

This secretive animal rarely shows its face above ground. But on a morning after a rain, you may catch a glimpse of it kicking dirt from a hole. A miniature volcano of soil gradually forms around the opening. Before long, the gerbil-sized kicker disappears again. A naked mole-rat is hard at work!

The unusual-looking naked mole-rats rest in their underground nest.

These volcanoes of soil are a sign that a naked mole-rat burrow system is underfoot.

Few people ever see naked mole-rats, but they know when the animals set up housekeeping. Small volcano-shaped mounds in fields, vacant lots, and dirt roads are a clue that naked mole-rats are busy underground.

Hidden below the surface is the naked mole-rat's burrow system. This maze of tunnels, containing nests and toilet chambers, may cover an area the size of 20 football fields. The underground city is home to a colony of naked mole-rats. Naked mole-rat colonies usually have 75–80 members, but sometimes as many as 300 individuals live together. They eat roots and dig new burrows with their razor-sharp teeth.

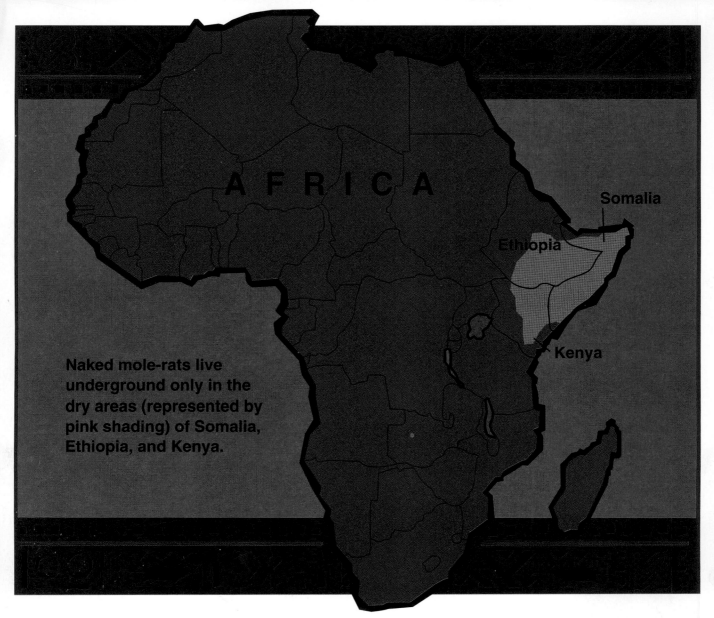

A F R I C A

Somalia

Ethiopia

Naked mole-rats live underground only in the dry areas (represented by pink shading) of Somalia, Ethiopia, and Kenya.

Kenya

Naked mole-rats are found only in the dry regions of Kenya, Ethiopia, and Somalia. These animals have been around for a long time. Fossils that look like skeletons of modern naked mole-rats have been discovered in 3-million-year-old rocks.

Kenyan farmers call naked mole-rats *uchi fuku.* In Swahili this means "naked mole." Although it lives underground like a mole and looks somewhat like a rat, the naked mole-rat is neither a mole nor a rat. Its skeleton and muscles are more like those of guinea pigs, porcupines, and chinchillas. These animals are the naked mole-rat's closest relatives.

Naked mole-rats are **rodents**—gnawing animals such as squirrels, mice, and hamsters. Naked mole-rats belong to the rodent family Bathyergidae. There are 12 species, or kinds, of mole-rats in this family. Only 1 species, the naked mole-rat, lacks a fur coat.

The naked mole-rat's scientific name is *Heterocephalus glaber*, which means "different-headed hairless." It certainly *does* look different from any other animal. But the mole-rat's odd appearance helps it to survive in its underground home.

The skeleton of a naked mole-rat

THE SECRET BURROW

The burrow system of a naked mole-rat colony may stretch for nearly 2 miles (3 km). The underground burrows are like our highways. Just as highways connect homes to shopping centers, the highway burrows connect nests to food areas, where roots grow.

The widest burrow is the main highway. It lies about 2 feet (50 cm) underground. This burrow is wide enough for two naked mole-rats to pass side by side. Along the way are turnaround spots where mole-rats can back in and change direction.

A naked mole-rat cleans up loose soil in a burrow after a cave-in.

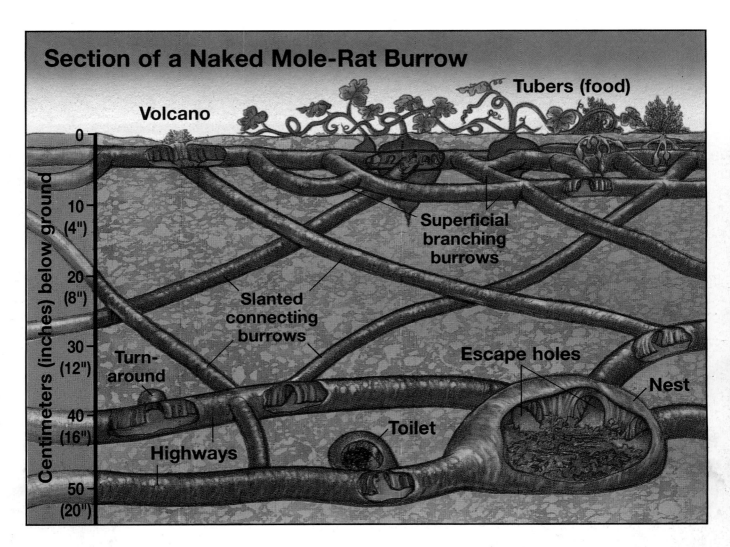

Section of a Naked Mole-Rat Burrow

Tubers (food)

Volcano

Superficial branching burrows

Slanted connecting burrows

Turn-around

Escape holes

Nest

Toilet

Highways

Centimeters (inches) below ground

0

10 (4")

20 (8")

30 (12")

40 (16")

50 (20")

Like side streets, narrower burrows branch off from the main highway. Some narrow burrows lead to food. Some lead to toilet chambers. Some are dead ends. Others lead to **volcanoes.** The burrows closest to the ground surface are so narrow that only one mole-rat at a time can pass through them. Most of these burrows lead to patches of small roots.

A mole-rat burrow system usually has many nests on the main highway. But the colony uses only the one or two nests that are closest to the best food source. When the food runs low, the mole-rats move to a new area and nest site.

Each nest chamber is about the size of a football. It is lined with grass rootlets and the skins of tubers. The entire colony snuggles together in the nest while sleeping or resting. Nests have several exits. If a hungry snake invades the

Above and Opposite: *Naked mole-rats pile on top of each other and snuggle together in their nest.*

burrow looking for a rodent meal, the mole-rats can escape out the back doors.

The toilet chamber is located in a dead-end burrow near the nest. Naked mole-rats urinate and defecate only in the toilet chamber. Since many animals live in close quarters, it is important to keep the rest of the burrow system clean. Disease might spread through the colony if wastes were not kept in one area. When the toilet chamber becomes full, the mole-rats dig a new one.

A BODY FOR
THE BURROW

The naked mole-rat, like all other **mammals,** has some body hair, gives birth to live young, and nourishes them with milk from the mother's body. But in many significant ways, the naked mole-rat is quite different from most mammals. Its bizarre body is made for underground life.

The naked mole-rat burrow is sealed, except when dirt is being kicked out of the volcano hole during digging. Little fresh air enters the burrow system. Many animals would suffocate in this stuffy environment because the air is low in oxygen. But not the naked mole-rat.

A naked mole-rat's loose skin and long body help it wiggle through narrow burrows and, . . .

Compared to other animals, the naked mole-rat can get more oxygen from each breath of air. In addition, it needs less oxygen to live than most animals its size.

Naked mole-rats control body temperature in a way that is unlike other nearly hairless mammals such as humans, whales, and elephants. These mammals keep the same body temperature even when the temperature of their surroundings changes. Their special body design helps them do this. Humans have sweat glands to help cool our bodies. A fat layer under our skin keeps in body heat. Like most mammals, we shiver to warm up. Whales have thick layers of fat that keep in heat. Elephants have thick, leathery skin as well as layers of fat.

...in this case, a large root through which it has just eaten.

13

The naked mole-rat has thin skin, no sweat glands, and no fat layer. It doesn't seem to shiver. Unlike other mammals, its body temperature changes with the temperature of its surroundings. In this way, naked mole-rats are more like reptiles, such as snakes and lizards, and amphibians, such as frogs and toads. So, how do naked mole-rats warm up and cool off?

The answer lies in the naked mole-rat's unique **habitat**. Its hot desert home has long dry seasons and two short wet seasons. Temperatures and rainfall above ground change. But the underground mole-rat burrows stay warm and moist year round. Since the burrow temperature remains a toasty 82–89° F (28–32° C) all the time, a naked mole-rat's body temperature doesn't vary much.

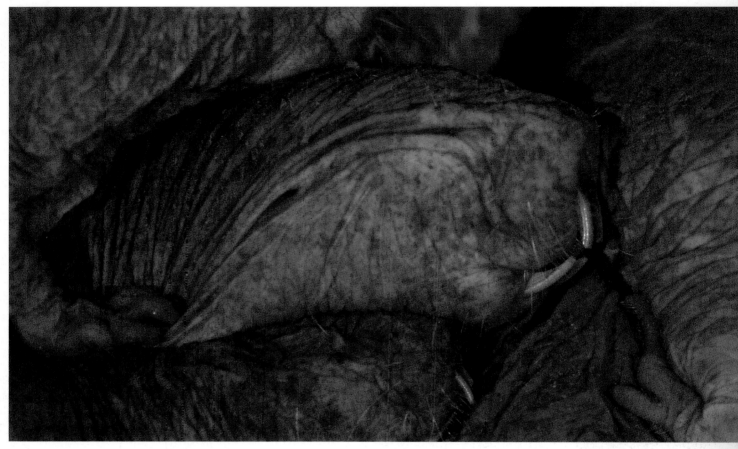

Naked mole-rats huddle for protection and warmth. Their thin wrinkly skin helps them heat up and cool down quickly.

In the cool early morning, a mole-rat warms up by lying in burrows near the ground's surface. It soaks up ground heat there in the same way that a snake warms up by lying on a sunny rock. After the mole-rat heats up, it hurries back to the nest and huddles with colony mates. The mole-rat's thin skin and lack of insulating fat help it act like a living hot-water bottle. Heat from its warmed body is transferred to its chilly colony mates through their thin, uninsulated skin.

When a mole-rat gets too hot, its thin, wrinkly skin also helps it cool off. The wrinkles and folds create more surface area for heat to leave the mole-rat's body.

Because its burrow home is always warm, the naked mole-rat doesn't need sweat glands for cooling or shivering for heating. Neither does it need fur, the way other mammals do, to insulate it from cold or heat.

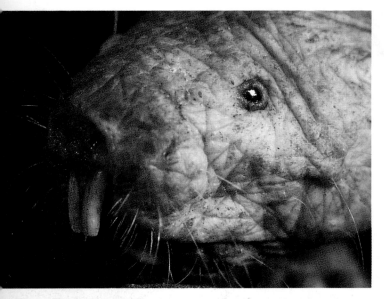

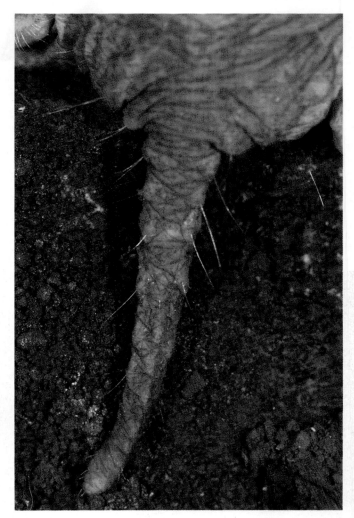

Naked mole-rats use their whiskers and the hairs on their tail to feel their way through dark burrows. The hairs between their toes are used to sweep up dirt.

Having fur all over would probably create problems for the naked mole-rat. Lice, mites, and fleas nest in mammals' hair and eat blood and skin cells. These tiny **parasites** can carry diseases, which would spread quickly through a crowded colony of furry animals.

Although the naked mole-rat isn't furry like most other mammals, it does have some hair on its body. It has a few whiskers on its cheeks and lips, scattered hairs on its body and tail, and tiny fringes of hair between its toes.

The mole-rat uses its few body hairs for touch the way people use fingers to feel their way down a dark hallway. As a mole-rat scurries through a burrow, its head swishes back and forth. Its whiskers touch the burrow walls and guide the animal. When the mole-rat moves backward, the hairs on its tail help find the way. A naked mole-rat can run backward as fast as it goes forward— something few other mammals can do.

Naked mole-rats usually travel through burrows with their tiny eyes closed. This probably protects the eyes from dirt and dust. Scientists think that the world looks fuzzy to a naked mole-rat. But since it lives in a dark underground home, the mole-rat doesn't need good vision.

Naked mole-rats travel through their burrows, often with their eyes closed.

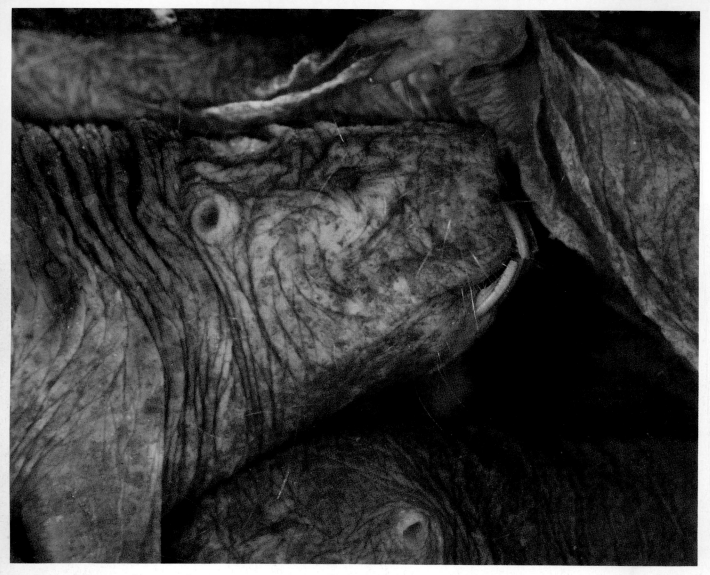

Naked mole-rats hear only a narrow range of sound. Because they live in burrows, naked mole-rats can only pinpoint sounds that come from in front of or behind them.

Other mammals have outer ear parts to direct sound to the ear. But naked mole-rats don't have outer ears. Mole-rats don't need to pinpoint sounds from many directions. That's because in their narrow burrows, sounds come only from behind or ahead of the animals.

Naked mole-rats are extremely sensitive to vibrations, though. Footsteps or digging on the ground above a burrow are enough to alert a mole-rat colony to possible danger.

A keen sense of smell helps naked mole-rats detect snake **predators** and other mole-rats invading their burrow. Body odor identifies an animal as friend or enemy. Colony members frequently roll around in their toilet chamber. Rolling in urine and feces keeps the colony's scent on the naked mole-rats' bodies. The scent is a way of saying, "I belong to this group." The animals also keep track of colony mates by nuzzling, shoving, tugging, and even nipping at each other.

A naked mole-rat rolls in a toilet chamber to refresh the colony's smell on its body. Naked mole-rats can roll halfway around inside their wrinkled skin.

Naked mole-rats chirp when they find food.

Naked mole-rats "talk" to colony mates, too. They make at least 18 vocal sounds, including grunts, chirps, squeaks, squeals, trills, and hisses. These calls announce danger, threats, anger, food, and the willingness to mate. No other known rodent produces so many types of sounds. That is probably because none lives with as many colony mates as the naked mole-rat.

POWER DIGGER

Scientists observed that one colony of 87 naked mole-rats had dug a 0.6 mile (1 km) burrow in less than three months—with only their teeth! Four sharp front teeth, called **incisors,** and strong jaw muscles make the naked mole-rat a super shovel. Hair-fringed lips close *behind* the teeth and keep soil out of the mole-rat's throat as it digs.

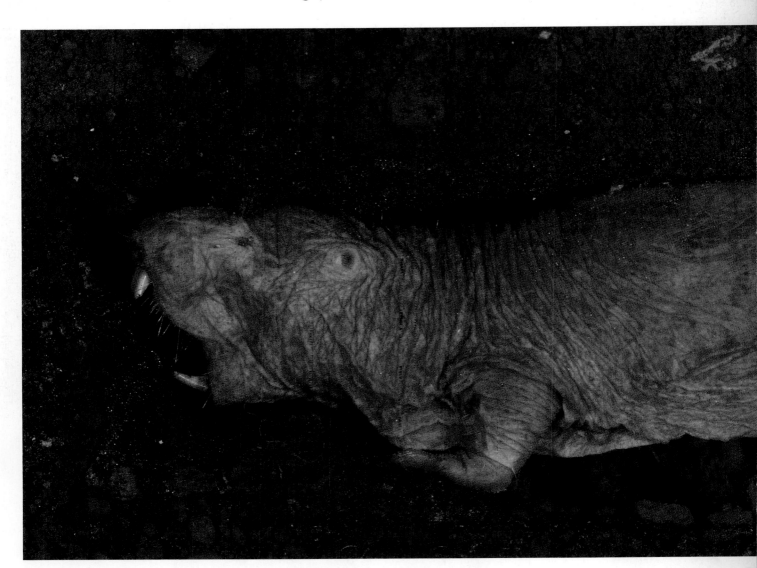

Strong wide-spreading jaws and sharp incisors help a naked mole-rat dig.

Mole-rats dig most of their new burrows during or right after the two rainy seasons, October–December and March–May. During the hot dry season, the fine clay soil is baked brick-hard. But the rains make the soil softer and easier to dig.

When it's time to make a new burrow, members of the mole-rat colony line up head to tail. Using its teeth, a digger mole-rat gnaws at the dirt. It collects a little pile of loose soil. Next the digger uses its small forepaws to claw the dirt under its body. Then it kicks the dirt backward with its hind feet.

The next mole-rat in line collects the pile of dirt. It scuttles backward in the burrow, using its back feet to sweep the dirt behind itself as it moves. The fringe of hair between its toes acts like a broom. The other mole-rats stand on tiptoe and let the sweeper pass beneath them.

A digger mole-rat in action

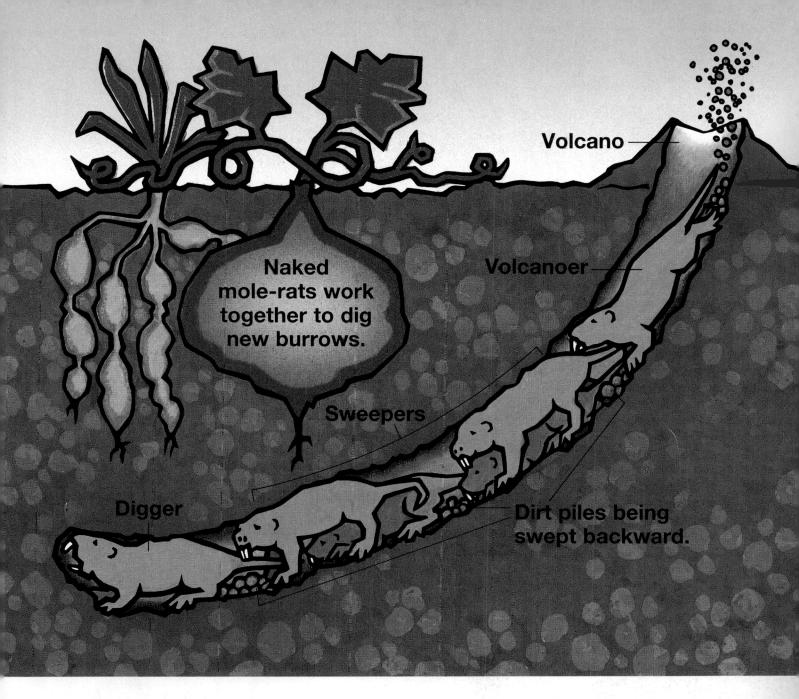

Volcano

Naked mole-rats work together to dig new burrows.

Volcanoer

Sweepers

Digger

Dirt piles being swept backward.

The sweeper stops near the surface opening of the burrow. There it kicks the dirt to a large mole-rat, called the **volcanoer.** The volcanoer kicks the soil out of the volcano hole. The sweeper returns to the front of the line for more dirt, walking on tiptoe and straddling the sweeping colony mates. The team of sweeper mole-rats forms a conveyor belt system of earth movers.

23

Dirt constantly reaches the volcanoer. This mole-rat kicks the soil out of the volcano hole in a continuous spray. The loose dirt forms a miniature volcano above ground about as tall as a ballpoint pen. When the job is done, the mole-rats seal the opening to the outside.

The main reason that mole-rats dig more burrows is to find food. Naked mole-rats eat underground plant parts such as roots, bulbs, and tubers. Bacteria, fungi, and special **protozoa** (microscopic animals) live in a naked mole-rat's stomach and intestines. These small organisms help the naked mole-rat break down and digest the tough, woody plant material.

A continuous spray of soil is kicked out of a volcano hole.

Underground plant parts like this large root are the mole-rat's food.

Naked mole-rats get both water and nourishment from the plants they eat. By munching on roots, a mole-rat can survive long, dry periods. Since the food is underground, it's protected from the heat and drought of the dry season.

Still, the plants grow far apart in the desert areas where naked mole-rats live. The animals have to do a lot of digging to find enough food to feed the colony. They seem to discover new food by luck. When they dig, they don't aim straight for roots. Sometimes they even dig right past a large tuber. It's likely that the mole-rats can't smell food because the hard soil doesn't carry odors well.

Some roots are big enough for the whole colony to share—and for a mole-rat to eat its way through!

In their blind search for food, naked mole-rats must gnaw through hard soil, often for long distances. A single animal would probably starve before it found a root. By living and working together, mole-rats can dig more burrows and explore larger areas. Then they have a better chance of finding a huge root—enough food for the whole colony to share.

AN AMAZING SOCIAL LIFE

A naked mole-rat colony is composed of a single **queen** and her colony mates. The queen controls the colony. She is the only female to give birth to the young and to provide them with milk. No other known **vertebrate** (animal with a backbone) lives in such a large colony with just one reproducing female.

A naked mole-rat colony, complete with a queen (lying on her back) and her three-day-old young, are piled up in their nest.

by its size. The small mole-rat on the left is too and a soldier is the large mole-rat on top.

Termites, ants, and some kinds of bees, wasps, and aphids are animals that live this way. Scientists once believed that the only eusocial animals were insects. That was before they found out about the naked mole-rat.

Like workers in an ant colony, each member of a naked mole-rat colony has a special job: housekeeper, soldier, or breeder. The **housekeepers** and **soldiers** include both males and females, but they do not breed. The queen and her mates do all the breeding. This system is called **reproductive division of labor.**

A mole-rat's job may change as it gets older. Young mole-rats begin to d housekeeping work when they are t

ess what a naked mole-rat does partly
k a housekeeper is in the middle

AN AMAZING
SOCIAL LIFE

A naked mole-rat colony is composed of a single **queen** and her colony mates. The queen controls the colony. She is the only female to give birth to the young and to provide them with milk. No other known **vertebrate** (animal with a backbone) lives in such a large colony with just one reproducing female.

A naked mole-rat colony, complete with a queen (lying on her back) and her three-day-old young, are piled up in their nest.

Naked mole-rats are eusocial animals, working together to keep their colony running smoothly.

Because of its special social organization, the naked mole-rat is considered a **eusocial** animal. Eusocial means "truly social." Eusocial animals live in family groups made up of parents and offspring. Only a few members of a group **breed** (produce young). The others work together to care for the young, provide food, and protect the group.

To get an idea of what a eusocial society would be like, imagine that your town is ruled by a queen. She is the only one who has children. Everyone else does a special job. Some people care for the queen's children. Some provide food for the entire town. Some protect the town from danger. Everyone in the town is related, but none of you ever has children of your own.

Termites, ants, and some kinds of bees, wasps, and aphids are animals that live this way. Scientists once believed that the only eusocial animals were insects. That was before they found out about the naked mole-rat.

Like workers in an ant colony, each member of a naked mole-rat colony has a special job: housekeeper, soldier, or breeder. The **housekeepers** and **soldiers** include both males and females, but they do not breed. The queen and her mates do all the breeding. This system is called **reproductive division of labor.**

A mole-rat's job may change as it gets older. Young mole-rats begin to do housekeeping work when they are two

You can guess what a naked mole-rat does partly by its size. The small mole-rat on the left is too young to work, a housekeeper is in the middle, and a soldier is the large mole-rat on top.

months old. Some will stay small housekeepers. Others will grow larger and become volcanoers or soldiers. A few will become **breeders.**

The housekeepers gather food, dig new burrows, and fix old ones. Using their teeth, they lift or drag chunks of dirt and small rocks out of the way. They keep the burrow walls smooth by gnawing off roots. They sweep loose soil to the volcanoer during burrow excavation. Housekeepers build nests by carrying grasses, rootlets, and skins of bulbs to the nest chamber. They bring food to the nest for others to eat and help the breeders care for the young.

This housekeeper is cleaning its burrow walls by biting off little roots that have grown into the burrow.

Sometimes naked mole-rats accidentally dig into another colony's burrow. Here soldiers from two colonies face off before fighting.

The soldiers are usually the largest nonbreeders. Their job is to guard the burrow. Bigger teeth and more powerful jaw muscles make them effective fighters. If the burrow system is invaded by a predator or by members of another mole-rat colony, the soldiers attack them.

Snakes are the main predators of naked mole-rats. Some snakes can smell freshly dug soil. When a volcanoer is kicking out dirt, a snake can slither into the volcano and attack the volcanoer. The other mole-rats quickly plug up the opening to the volcano with dirt. This prevents the snake from entering the burrow. If a snake does sneak into a burrow, it is bitten by soldiers.

The poisonous rufous-beaked snake is one of the naked mole-rat's enemies.

The queen is the only female breeder. She chooses one to three males as her mates. They are usually closely related to her and may even be her brothers. The male breeders may have been house-keepers or soldiers when they were younger. But once they are chosen to be the queen's mates, their only job is to breed and help care for the pups.

The queen doesn't have to search for

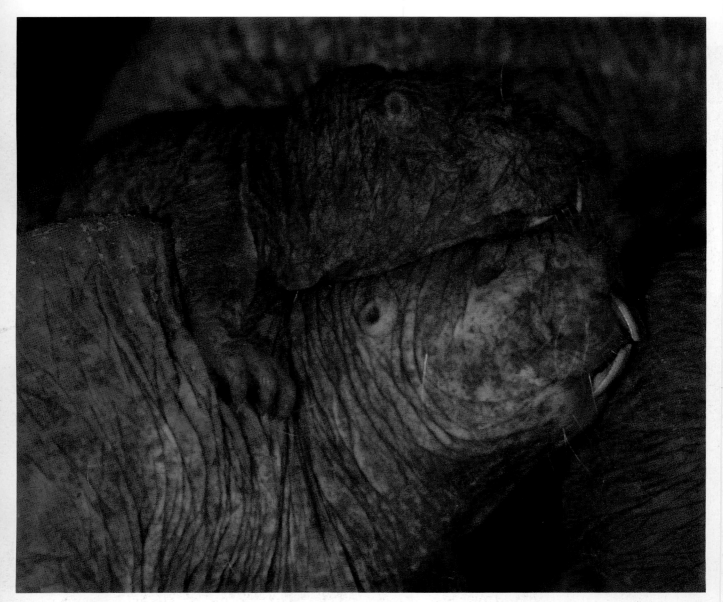

The queen (bottom) is much larger than the other mole-rats in the colony and is the only female breeder.

food, build a nest, or defend her young. The housekeepers and soldiers do that for her. As a result of this "royal treatment," the queen and her mates can produce many offspring. The breeders successfully rear four to five litters a year, each containing about 10–15 pups. Unlike most other mammals, naked mole-rats breed during all seasons, probably because food is available year round.

Naked mole-rat queens have up to five litters a year.

THE BOSSY LADY

In a naked mole-rat colony, the queen runs the show. She spends much of her time patrolling the burrows. When she sees work to be done, she uses her nose to shove others into action.

If the queen sees a cave-in, she shoves a housekeeper toward the job site. If she senses danger, she shoves the soldiers to defend the colony. If she realizes that food is needed, she shoves housekeepers out to find a tasty root. Because she keeps workers busy, the colony stays fed and the burrows are expanded and defended.

A queen hisses and shoves a worker to do its job.

A worker freezes in response to a queen's aggression. When the queen leaves, the worker will scamper off to do its job.

The queen mole-rat bosses some individuals more than others. She rarely shoves her brothers or sisters or her offspring. But she bullies larger members of the colony, especially if they are not closely related to her. These distant relatives are the ones most likely to challenge her power. They are also the laziest in the colony. Her shoving and pushing keeps them working. By bullying the other females, the queen also stops them from reproducing. Her control over the colony is powerful.

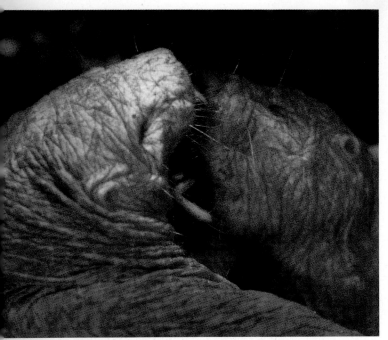

When the queen becomes weak or dies, the battle to replace her begins. Many other adult mole-rats, male and female, gain weight. The largest females fight each other by shoving, biting, or fencing with their front teeth. The battles may go on for weeks or months. Finally one female wins by crippling, killing, or frightening the others.

Large females fight to be the next queen.

A queen huddles with her colony mates a day before giving birth to 27 pups. She had more than 900 pups in her 12-year lifetime!

After the winner takes control of the colony, her body gets longer. She ends up one-third longer than her colony mates and usually heavier, too. No other known mammal increases in length *after* reaching adulthood. The added length allows space for a large litter of pups to develop without the queen getting too wide to fit through the burrows.

THE NEXT GENERATION

The queen's pregnancy lasts 10–11 weeks. Shortly before the pups are born, the entire colony huddles with the queen in the nest. After the pups are born, nonbreeders lick them clean. Then the queen begins to nurse her babies.

Like the adults, the newborn pups have no fur. But they do have cheek whiskers and scattered body hairs used for touch. The pups are able to crawl at birth. Even though they are blind, they can sense their mother, probably by smell. When she enters the nest, they wobble over to her to nurse.

The other members of the colony huddle with the pups in the nest and keep them warm. The queen, the male breeders, and the younger nonbreeders lick and groom the pups. If the burrow is threatened, the older animals use their teeth to grasp the pups by the neck or belly. Then they carry the babies to a different nest away from the danger.

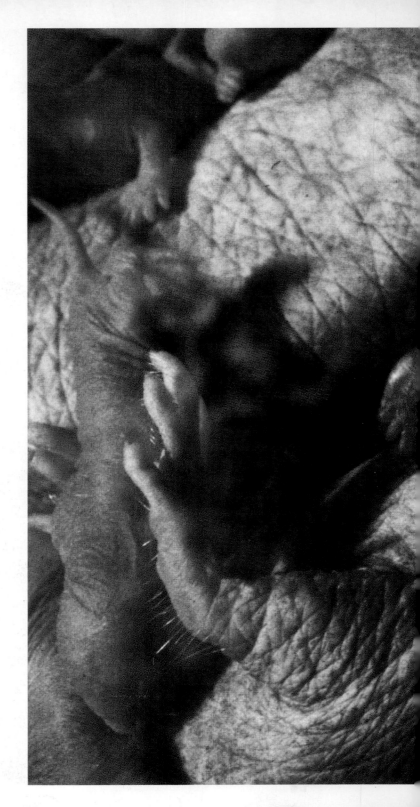

Newborn pups sleep and nurse. Older pups also practice being adults by wrestling and fencing with their teeth.

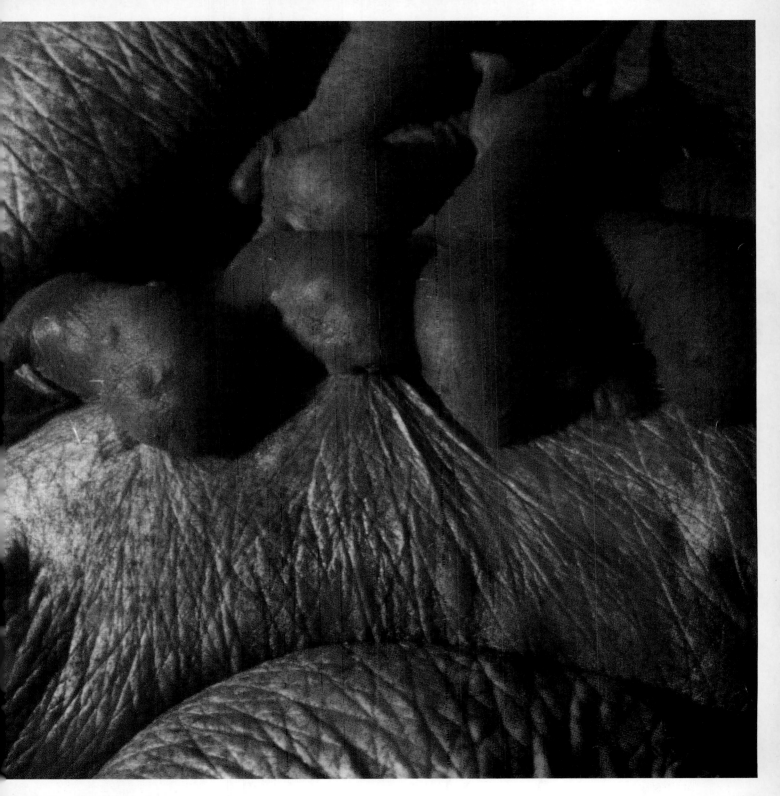

41

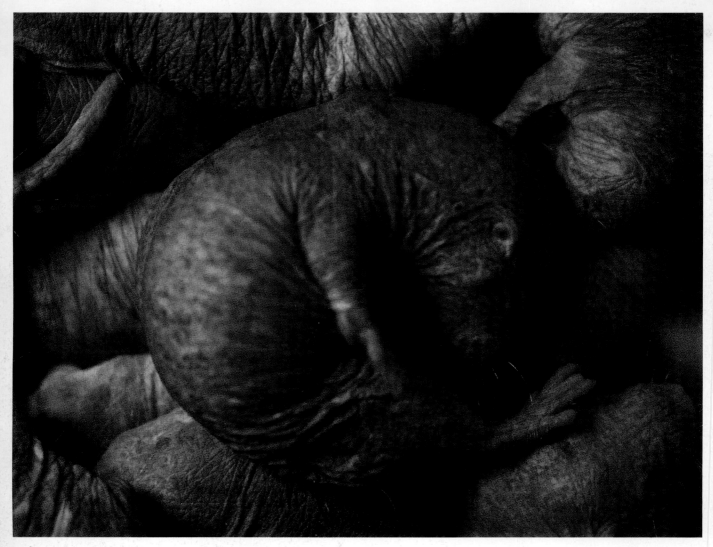

A naked mole-rat eats its own cecotrophes.

After two or three weeks, the pups begin to crawl into the burrows and feed themselves. **Cecotrophes** are their first solid food. These are soft, green waste pellets produced by older colony members. Naked mole-rats produce two kinds of pellets. One is pure waste, and the other, cecotrophes, is an important source of nutrition. Cecotrophes contain the tiny organisms (bacteria, fungi, and protozoa) that will break down tough plant material when young naked mole-rats begin eating roots. The pups beg for cecotrophes by making a mewing call and nuzzling the adult's anal region.

After one month, the pups stop nursing and begin to eat only solid foods. Their eyes open and they grow rapidly. When they are two months old, the young mole-rats start work as housekeepers. They will find food and carry it back to the nest for their parents and the next litter of pups.

A young mole-rat's first job is that of housekeeper. One of its jobs is to carry food to the nest.

A CREATURE LIKE
NO OTHER

The naked mole-rat is an unusual animal. It gives birth and nurses its babies the way mammals do. It controls its body temperature the way reptiles do. It lives in large colonies with a single reproducing female the way social insects do.

But thanks to its unique body and remarkable social life, the naked mole-rat survives where other animals cannot. In its desert home, the soil is dry and hard. Food is widely spread out. By living in a family group instead of alone, naked mole-rats have a better chance of surviving. Working together, colony members can dig tunnels, find food, and fight enemies. All of this would be difficult for a single animal to do.

Although naked mole-rats may be considered ugly, their bodies are perfect for their underground lifestyle.

This eusocial lifestyle has attracted the attention of many scientists. Some researchers have traveled to Africa to study how the naked mole-rat lives in its natural home. Others have brought colonies into laboratories where they can watch the animals closely. The researchers' work has led to the knowledge we now have about naked mole-rats. By continuing to study mole-rats, scientists hope to understand more about why many animals live together and how they get along.

You can meet a naked mole-rat, too. Zoos all over the world, including many in the United States, display naked mole-rat colonies. Visit with these extraordinary creatures and watch them in action. Even though they won't win any beauty contests, you won't be able to keep your eyes off them.

Most naked mole-rats live 2 or 3 years in the wild. Queens live twice as long. In laboratories, naked mole-rats have lived 20 years.

GLOSSARY

breed: to produce young

breeders: animals that produce young

cecotrophes: soft partly digested pellets produced by adults and eaten by both adult and young naked mole-rats

eusocial: truly social. Living in a large family group in which only a few members produce all the young, and the other members cooperate to defend and maintain the colony

habitat: the area in which an animal normally lives

housekeepers: members of a naked mole-rat colony that keep burrows clean, build nests, and gather food

incisors: large, sharp front teeth

mammals: animals with hair on their body, that give birth to live young and feed their young with mother's milk

parasites: an animal that benefits from living with, in, or on another animal and usually causing it harm

predators: animals that hunt other animals for food

protozoa: tiny, microscopic animals

queen: the female breeder and ruler of a naked mole-rat colony

reproductive division of labor: a social system in which only a few individuals in a group produce young

rodent: the scientific order of gnawing mammals that includes naked mole-rats

soldiers: large members of a naked mole-rat colony that fight off invaders

vertebrate: animal with a backbone

volcanoer: a large member of a naked mole-rat colony that kicks soil out of the burrow

volcanoes: above ground mounds of soil resulting from burrow digging by naked mole-rats

INDEX

ABOUT THE AUTHORS

Gail Jarrow is the author of several novels for young readers as well as numerous magazine articles and stories. Her interest in rodents began when she raised guinea pigs as a child. After studying zoology at Duke University, Gail taught science and math in elementary and middle schools, where her classrooms were home to mice, white rats, and other animals. She now lives with her husband and three children in rural New York State. Paul Sherman is her next-door neighbor. As a result of their friendship, they decided to combine their knowledge and skills in writing this book about naked mole-rats.

Paul Sherman received his undergraduate degree in biology at Stanford University and went on to receive his Ph.D. in zoology at the University of Michigan, in Ann Arbor. He first became interested in naked mole-rats as a postdoctoral fellow at the University of California, Berkeley. Now Paul is a professor of animal behavior at Cornell University. He continues to study the social behaviors of various birds and mammals, including naked mole-rats, Idaho ground squirrels, and wood ducks. Paul describes naked mole-rats as one of the most important recent finds in animal social behavior.